Fuels and Petrochemicals Division 2021

Held at the 2021 AIChE Spring Meeting and 17th Global Congress on Process Safety

Online
18 - 23 April 2021

ISBN: 978-1-7138-5342-8

Printed from e-media with permission by:

Curran Associates, Inc.
57 Morehouse Lane
Red Hook, NY 12571

AIChE
120 Wall Street, FL 23
New York, NY 10005-4020

Phone: (800) 242-4363
Fax: (203) 775-5177

www.aiche.org

Additional copies of this publication are available from:

Curran Associates, Inc.
57 Morehouse Lane
Red Hook, NY 12571 USA
Phone: 845-758-0400
Fax: 845-758-2633
Email: curran@proceedings.com
Web: www.proceedings.com

Fuels and Petrochemicals Division 2021

Held at the 2021 AIChE Spring Meeting and 17th Global Congress on Process Safety

Online
18 - 23 April 2021

AIChE
The Global Home of Chemical Engineers

TABLE OF CONTENTS

620660 The Digital Transformation Promises to Change Innovation in the Materials Industry. Will It?

Abstract:

The industry is challenged to innovate faster than ever while also delivering sustainability improvements that require profound change through the value chain. Simultaneously with that pressure, digitalization and AI are advertised as revolutions that has caused many to fear becoming obsolete. This dichotomy of extreme pressure and emerging, revolutionary technologies is causing fast, significant change in the industry and the innovation process. In this talk, I would like to share my view on what this impact will be, what risks exist, and what, in my view, will drive long-term success.

620661 Digital Twins & Smart Manufacturing –*Where Data Analytics Meet Modeling & Optimization*

Abstract:

Modeling and Optimization constitute core competencies of process systems and chemical engineering fields, and play pivotal role in the analysis, simulation, design and operational activities that we perform daily in order to improve the performance and sustainability of energy and manufacturing processes and supply chains. The re-emergence of AI and Big Data revolution that we are all witnessing at the moment, also provide an opportunity and a platform toward real-time decision-making. While often perceived as different fields, in the context of Smart Manufacturing, Data Analytics and Modeling/Optimization can in fact be viewed as 'two sides of the same coin'.

In this lecture, we will attempt to elucidate this argument through the lenses of ***hybrid modeling and digital twins***, which bring together the advantages and power of high-fidelity modeling & optimization and AI/machine learning technologies and tools, and their synergy. A central theme will be that the evolution/revolution in data-driven methods in the context of process modeling and automation is becoming a critical component in advancing the energy & manufacturing transition agenda toward clean energy, increased process/energy efficiency, sustainability and resilience, and ultimately economic and societal prosperity.

621233 Color Changing Processes: Moving from Black to Green

Abstract:

Major changes are coming to the Fuels and Petrochemicals' industries. In the drive to become less carbon intensive, producers are looking at integrating Blue, Green or Circular economies to existing facilities. In the drive to becoming less carbon intensive, facilities will need to investigate their process and determine what changes they can make without impacting their overall production or product quality. Some questions that arise during this change could be: How can existing facilities integrate Green or Blue methods into their processing, such as Hydrogen? Does this change make our process "Green" or "Blue" or are we shifting the "Black" to someone else? How can we maintain availability with unpredictable energy sources?

619099 APC Toolkit: Choosing the Right Tool for the Job

Abstract:

The term Advanced Process Control (APC) is generally associated with complex model-based, multivariable control (MVC) applications. However, the word “advanced” simply refers to something that is "beyond the ordinary or introductory". Significant process control improvements can often be achieved by implementing advanced regulatory controls (e.g. cascade, ratio, feed forward controls), utilizing different PID algorithms and features available in modern control systems, and/or incorporating engineering knowledge through the use of linearization techniques such as valve characterization or gain scheduling. Higher level APC solutions, such as MVC, are usually more appropriate for processes with complex interactions, multiple constraints and/or multiple objectives. Most successful APC applications include an optimal combination of some or all of the above. This presentation will review many of the tools and techniques commonly used in APC applications, including some practical examples and recommendations for choosing the right tool(s) for the job.

Abstract

Multifaceted Biochar: biofuel source, graphene precursor and methane-to-hydrogen catalyst
Presenting author: Rahul Kundu, email: rak628@utulsa.edu
Advisor: Dr. Hema Ramsurn
Russell School of Chemical Engineering, The University of Tulsa, OK, 74104

Overview:

Fossil fuels have a finite reserve and if other carbon sources are not explored, our future global energy demands will not be met. Biomass represents a viable solution due to its abundance and accessibility. Although various techniques, such as pyrolysis and hydrothermal liquefaction and gasification, have been explored to convert biomass to biofuels, there are various challenges associated with these methods. One of the main problems is the formation of biochar and tar as byproducts which reduce the overall conversion of biomass to value-added products. To make the bio-refinery concept more sustainable, any byproduct like char must be converted to valuable products such as biofuel and/or other carbon-based materials. Therefore, my research is focused on the **synthesis of value-added products from biochar obtained by hydrothermal carbonization of biomass model compounds (cellulose and lignin)**. The main advantage of studying biomass model compounds is to better understand their behavior separately as cellulose and lignin are structurally different and consequently translate that to biomass. In my research, I prepared biochar by hydrothermal carbonization in a batch reactor (220-300°C and 3.5-10.3 MPa). Using this biochar as a raw material, I have worked on synthesizing three different products: (a) biofuel, (b) carbon-based nanomaterial (graphene) on metals, and (c) carbon encapsulated iron nanomaterials for methane to hydrogen conversion. During my research I have designed my experimental set-up and gained experience in various analytical instruments including GC-MS, FTIR, Raman Spectroscopy, BET surface area analysis, TEM, XPS and SEM.

Research Experience:

For the first part of my research, I used biochar as source to produce biofuel using tetralin as a hydrogen donor solvent at 400°C. I observed that the conversion of cellulose biochar was higher than lignin biochar because cellulose and lignin are structurally different, and lignin is harder to breakdown. Using analytical techniques such as FTIR and GC-MS, I was able to understand and postulate a mechanism for biochar dissolution in tetralin. I concluded that the dissolution of biochar was higher when the C/O of biochar was low and in the presence of biocrude (produced during hydrothermal carbonization). Further, I studied the kinetics of cellulose biochar dissolution where the products formed at various temperatures during the heating step of the dissolution process were analyzed and the major components in the products such as naphthalene (a byproduct from tetralin), levulinic acid (obtained from cellulose) etc. were quantified using GC-MS to understand the role of tetralin during dissolution. I observed that biochar dissolution is a two-step process in the heating stage where first the unconverted biomass in the biochar decomposed followed by the dissolution of biochar. Based on the experimental data, I proposed a new temperature dependent kinetic model where the activation energies were evaluated by non-linear regression. Since, all the conversion occurred during the heating stage of dissolution reaction, the activation energies were evaluated in the heating stage where the rate constant was function of temperature according to Arrhenius equation.

In recent years, graphene, a carbon-based nanomaterial has gained popularity among researchers. Graphene has been called a "wonder material" due to its unique properties. It was discovered by Dr. Novoselov and Dr. Geim in 2004 who won the Nobel prize in 2010. Graphene is a single layer of carbon (sp^2) which could be obtained by exfoliation of graphite. It has the potential to be used in biosensors to detect disease like cancer, in microprocessors, in batteries etc. To produce high purity graphene, chemical vapor deposition (CVD) has been used to produce flakes (up to 100 μm) of graphene on copper. Although CVD produces high purity graphene, this method requires greenhouse gas like methane as a source of carbon along with H_2 (flammable) at very high temperature. Also, in most cases CVD requires vacuum in the reactor. These constraints make the process expensive and dangerous. Therefore, to mitigate these issues, I worked on developing a process to use biochar as a source of carbon to produce few-layer graphene on metal foils at atmospheric pressure which does not involve the use of hydrocarbon gas and hydrogen. This makes the process simpler, safer, environment friendly and most importantly less expensive and scalable. While working on this project, I have learned how to use various analytical techniques to characterize a nanomaterial such as scanning electron microscopy (SEM with Energy Dispersive X-Ray Spectroscopy), X-Ray photoelectron spectroscopy (XPS), Raman spectroscopy and X-Ray Diffraction (XRD). The Raman spectroscopy analysis revealed that the presence of the few-layer graphene on all the foils where the 2D/G ratio was less than 1. XPS showed over 80% purity (sp^2 carbon) on the surface of the foil. This innovative process resulted in a patent for which I am one of the inventors (patent's PCT publication number: WO2020/086841). I helped in tailoring our scientific invention into legal terms by working together with my advisor and our legal and IP team. This innovation led me to participate in the i2e Love's Entrepreneur's Cup Business Competition where I teamed up with 2 students from University of Tulsa business school and competed against 30+ teams. Our team won the interview round and was among the 6 selected teams which qualified for the finals. In this competition, we formulated a business plan to produce graphene using our new process mentioned in large scale.

Industrial and technological developments have resulted in rapid increase in the rate of global energy consumption with the energy demand expected to increase to 33% by 2035. Fossil fuel resources are the dominant source of energy and during 2018, the oil demand grew by 1.5% while that of natural gas grew by 5.3% which is the fastest rate of growth since 1984. Due to the increased use of fossil fuels, the amount of carbon emissions has grown by 2%, the highest in the past seven years. This also led to an increase in emission of greenhouse gases (GHGs) like CO_2, NO_x and SO_x which contribute to climate change. Unfortunately, with the current technologies, an efficient and economical route for utilizing renewables including solar, bio and wind as a source for large scale commercial energy production is not yet available. This gives significant scope for hydrogen, a clean source of fuel to be the future source of energy. Combustion of hydrogen produces only water and advances in hydrogen fuel cell technology can significantly decrease the GHG emissions. Compared to liquid hydrocarbons (13.1 KWh/kg), hydrogen (39.4 KWh/kg) can produce three times higher chemical energy per unit mass. The thermal decomposition of methane is the only single step process which converts methane to hydrogen without producing CO_x. Therefore, I am currently working on synthesizing biochar-based carbon encapsulated nano sized metal (Fe) catalyst from biochar to produce hydrogen from methane in a lab scale continuous flow reactor system. Characterization techniques such as XRD (to study the crystalline structure of the catalyst), BET surface area (to determine the surface area per gram using N_2), SEM, HRTEM (high resolution transmission electron microscopy) etc. are being used to study the synthesized catalysts. The catalyst is then used in the conversion of methane to hydrogen in a continuous flow reactor.

Teaching and Mentoring Experience:

Apart from my research experience, I was also a teaching assistant from August 2017 through May 2020. My responsibilities included managing undergraduate chemical engineering laboratory and grading homework and quizzes (fluid mechanics, heat transfer and thermodynamics). In the undergraduate lab, I assisted students in performing laboratory experiments related to enzyme kinetics, composition analysis, bomb calorimeter, distillation, packed tower, viscometer, and pump performance. I also had the opportunity to mentor undergraduate and high school researchers.

Scholarly Achievements:

- Peer Reviewed Journal articles: 2 published (first authored)
- Patent: 1
- Conference Presentations: 7
- Bellwether Fellowship, University of Tulsa (May 2020- May 2021)
- Grant from Office of Research and Sponsored Programs, University of Tulsa (2019)
- Won the interview round and finalist in the i2e Love's Entrepreneur's Cup Business Competition (April 2019)
- Nova Day of Innovation Award (University of Tulsa, 2019)
- Chapman Scholar Presentation Award (2018)
- Phillips 66 R&D Fellowship (Summer 2017, 2018 and 2019)
- Graduate Research and Teaching Assistantship (2016-Present)
- Grant from Student Research Grant Program, The University of Tulsa (2016 and 2020)

Peer Reviewed Publications and patent:

- **Kundu, R.**; Ramsurn, H. "***Non-catalytic Dissolution of Biochar in Hydrogen Donor Solvent***" Journal of Analytical and Applied Pyrolysis, 140 (2019): 227-238. DOI: 10.1016/j.jaap.2019.03.020
- **Kundu, R.**; Ramsurn, H. "***Kinetic study of non-catalytic dissolution of cellulose biochar in hydrogen donor solvent***" ACS Sustainable Chemistry & Engineering, (2020): DOI: 10.1021/acssuschemeng.0c02907
- **PATENT:** Vacuum-Free, Hydrogen-Free Catalytic Synthesis of Graphene from solid hydrocarbons **(PCT publication number: WO2020/086841)**

619291 Integrated Scheduling for Petroleum Refinery Supply-Chain Management

Abstract:

The petroleum refinery supply-chain is a leading segment over the oil industry-wide supply-chain, covering the plant-wide petroleum refinery scheduling sub problem and the multi oil-product pipeline distribution sub problem. These sectors along the petroleum refinery supply-chain are highly correlated, since the output of refinery processing sub system is the input to the product distribution sub system. The oil companies are aiming at maximizing the profit margin by weighing benefits versus costs throughout the entire refinery supply-chain. According to Chima (2007), there is a need to ensure that each sector along the supply-chain can respond quickly to the exact demand of its downstream customers, protecting itself from problems with suppliers and buffering its operations from the demand and supply uncertainty it faces. Thus, it can greatly maximize the potential benefit margin by coordinating the management and operations of the two sub problems along the refinery supply-chain. There exists a lot of works explicitly and deeply exploited and studied these sectors separately, while only a few works addressed the sectors along the petroleum refinery supply-chain simultaneously. Guyonnet et al. (2009) integrated the crude oil unloading, production planning, and distribution sub models on a tactical decision level by solving each part in a sequential push or pull manner, where the planning horizon is discretized into time periods, typically a day or a week. Siwi et al. (2018) proposed a strategical planning of the petroleum and petrochemical supply chain. However, systematic studies for the integrated scheduling of these sectors are still lacking.

This work is aiming at realizing the holistic scheduling and optimization of the petroleum refinery supply-chain. Crude oil unloading, transferring, manufacturing, and oil-product distribution systems are integrated optimized. The scheduling objective is to minimize the total operating cost; meanwhile, operating rules (including newly considered operating rules in Yu et al. (2020), such as tank receiving and discharging operations cannot be performed simultaneously, which can further close the gap between the modeling and the realistic problem), and requirement specifications for crude oil blending and oil product blending (where nonlinearity may occurs due to the blending operation), inventory limits, delivery feasibility constraints and oil product demands must be satisfied, which may result in a large-scale MINLP model. The efficacy of the developed integrated scheduling model is demonstrated by an industrial-scale case study.

Keywords: Integrated scheduling, MINLP, Petroleum refinery supply-chain, Crude unloading, Refinery manufacturing, Multi oil-product, Pipeline distribution

References

Guyonnet, P., Grant, F.H., Bagajewicz, M.J. (2009). Integrated Model for Refinery Planning, Oil Procuring, and Product Distribution. Industrial & Engineering Chemistry Research, 48: 463–482

Siwi, R. G., Aljumah, F., Li, J., & Xao, X. (2018). Optimal Strategic Planning of Integrated Petroleum and Petrochemical Supply Chain. In Computer Aided Chemical Engineering (Vol. 43, pp. 1201-1206). Elsevier.

Tong, K., Feng, Y., Rong, G. (2012). Integrated Model for Refinery Production and Pipeline System Scheduling. Proceedings of the 22nd European Symposium on Computer Aided Process Engineering,

17 - 20 June 2012, London.

Yu, L., Chen, M., & Xu, Q. (2020). Simultaneous Scheduling of Multi-product Pipeline Distribution and Depot Inventory Management for Petroleum Refineries. Chemical Engineering Science, 115618.

619212 Heat Release Analysis of Neat and Upgraded Biorenewable Thermal Deoxygenation Oils in a Diesel Engine

Abstract:

Thermal DeOxygenation is a non-catalytic thermochemical process which converts biomass-derived levulinic and formic acids to crude hydrocarbon oils at 450°C and atmospheric pressure. The resulting oils contain a broad boiling point distribution of hydrocarbons (75-550°C) with low oxygen content (<5 wt%) and low total acid number (TAN) compared to pyrolysis oils. Crude oil distillates (150-325°C) have been analyzed as an ultra-low sulfur diesel (ULSD) fuel blend stock at 50 vol% in a naturally-aspirated diesel engine. High aromaticity and low cetane rating of the blend delayed combustion phasing and resulted in high specific engine emissions of CO and THC compared to USLD. Hydrogenation and distillation of the crude oils over Ni/SiO_2-Al_2O_3 catalyst at 350°C improves cetane and hydrogen content of the resulting diesel fraction to meet most American Society for Testing and Materials (ASTM) D975 #2 distillate fuel specifications.

In this study, crude TDO oil is hydrogenated using a commercial Ni/SiO_2-Al_2O_3 catalyst in bench scale continuous flow tubular reactors. The crude and hydrogenated TDO oils are distilled to produce jet fuel (180-250 °C) and diesel (150-325 °C) range hydrocarbons using a fractional distillation apparatus according to ASTM standards. Four different TDO oil blendstocks (crude TDO oil, Hydrotreated TDO oil, distilled TDO oil, and distilled Hydrotreated TDO oil) are analyzed for chemical and physicochemical properties following ASTM standard methods and used for diesel combustion analysis. Properties of the TDO fuels are compared with ASTM and military fuel standards (MIL-DTL), and blended appropriately to meet the specifications. Each TDO diesel blendstocks is blended with ULSD in four different volume percent (5 %, 10 %, 15%, and 20%) to produce sixteen total fuel blends for the analysis. Combustion and emission behaviors of the fuel blends were studied using a naturally-aspirated, air-cooled, single-cylinder diesel engine as a function of engine load at a fixed speed of 3,600 RPM. In-cylinder heat release analysis results show combustion phasing and indicated thermal efficiency can be predicted by fuel cetane rating. Quantitation of engine emission include oxygenates (aldehydes and ketones) using GCMS methods and soot which is collected on 0.45 um fiberglass filters and analyzed using thermogravimetric analysis to ascertain total emission rate and volatile fraction. Results show TDO oils are a viable renewable fuel blending agent for making finished transportation fuels.

618851 "Development of a Production Scale CO2 Conversion Catalyst for a Single System Circulating Fluidized Bed to Create Value Added Chemicals"

Abstract:

Kyle March, Albert Zwiener, Chemical Engineering Dept., Division of Chemistry and Chemical Engineering, Southwest Research Institute, San Antonio, TX 78238.

Topic: Fluidized reactor technology /Carbon capture/CO2 utilization

"Development of a production scale CO2 conversion catalyst for a single system circulating fluidized bed to create value added chemicals"

Current carbon dioxide emission reduction plans focus on mitigating emissions at the industrial process point source. Typically capture and separation process for carbon dioxide create a concentrated stream of carbon dioxide that may be used as a feedstock for chemicals or fuels. SwRI has been on the forefront of this effort for more than 10 years, and has continued to search and work with its clients for innovate ways of capturing, separating, and utilizing carbon dioxide. SwRI is working to develop scaled up production catalyst and processes to convert carbon dioxide to value added products using temperature and pressure in a circulating fluidized bed reactor. By combining SwRI's knowledge and experience with circulating fluidized beds, catalyst development, and carbon dioxide conversion, a process for converting carbon dioxide to value added products is being researched. This presentation will discuss the development of a CO_2 conversion catalyst from lab scale testing and production to larger-scale production technique as well as its viability in a circulating fluidized bed.

618796 Novel Techniques to Improve Polymer Reactor Temperature Controller Tuning

Abstract:

Good control of polymer loop reactor temperature improves product quality, reduces product transition time and increases production rate. However, due to the sensitive nature of the exothermic process, and an operating philosophy that is sensitive to any potential process disturbances during loop testing, it is difficult to make improvements to the tuning of the control loops.
Measuring process dynamics on a polymer loop reactor and related cooling loops can be difficult due to the highly exothermic reaction and different operating conditions for the many products. Combine these factors with an operating philosophy that is sensitive to any potential process disturbances during loop testing and it is difficult to measure process dynamics and improve loop tuning. Even minor deviations from the reactor temperature set point can result in a deactivation of the reactor by the safety system.

In this case. a combination of technologies was used to improve reactor and reactor coolant temperature loops. First, the process dynamics were approximated from closed loop response of the coolant loop and reactor temperature during normal operation. The coolant and reactor temperature processes were simulated within native control system. The simulation was used to evaluate the improvement from improved loop tuning and to develop an open loop step test plan that would satisfy the concerns of the operation department. Then, open loop step tests were performed to obtain more accurate process dynamics which were used to improve the accuracy of the simulation. The improved simulation was used to further improve the reactor tuning and control performance.

The improved tuning and control scheme changes reduced the reactor temperature variability, reduced the impact of heat exchanger swaps, reduced product transition time and reactor startup time. These changes resulted significant increase in production capacity.

617452 The Power of RAM Modeling: Optimizing Facility Performance throughout Life

Abstract:

Projects in the energy industry often involve the design and construction of complex facilities, necessitating upfront capital investment and ongoing operational expenditure. Decisions made during design can carry considerable risk, both in terms of future profitability, as well as the impact that incidents may have on health, safety and the environment.

This presentation will introduce the audience to Reliability, Availability and Maintainability (RAM) system design attributes, considering the use of RAM Modeling to simulate the varying effects of equipment failures over facility lifetimes to support statistically relevant predictions of facility throughput. Simulations can be used to identify the main bottlenecks to production and help management formulate a targeted improvement plan. Proposed improvements also bring with them the need to determine the impact on risk, so the necessity to have the reliability engineer and the risk assessor working together will be highlighted in this presentation.

619130 Sustain and Improve APC Performance without Breaking the Bank

Abstract:

Since most Advanced Process Control (APC) applications are model-based, they will require on-going maintenance to sustain their performance and benefits. Poor APC performance is often due to a difference (mismatch) between the APC model and the actual process behavior. Model mismatch may be the result of normal changes in process behavior over time (e.g. fouling), minor process improvement projects (e.g. replacing a heat exchanger or valve), major debottleneck or turnaround projects, or other operational changes. An effective APC maintenance plan requires close coordination with plant operations, maintenance, planning/scheduling, project engineering, etc.

Making the changes required to improve the performance of a complex APC application can be very intimidating, In some extreme cases, there may even be a reluctance to make process improvements that might impact a critical APC application. APC should never be an obstacle for process improvement projects (or vice versa). Maintaining an APC application is often thought of as a major project that may involve significant costs and/or disruption to operations. But with preliminary planning and analysis, maintaining an APC application isn't as costly or disruptive as many people believe.

This presentation will review general tools and techniques that can be used to identify the source and scope of APC performance issues and several different options if the model needs to be updated. In many cases, standard trending and data analysis tools can identify specific instrumentation, regulatory control and/or APC issues. There are several approaches to consider if the model needs to be updated: manual model adjustments (stop-gap solution), focused testing (when only a small part of the APC is affected), closed-loop testing (when the model mismatch is relatively small). In some cases, dynamic and/or steady-state simulation can be used to obtain or validate a model, with an appropriate degree of caution and verification.

617526 7 Ways to Monitor Controller Variability – Which One Is Right for You?

Abstract:

Control systems exist, primarily, to respond to upsets and maintain the process at its operating point. In other words, to stabilize the plant and minimize variability. Millions of dollars are invested in instruments, controls, pumps, motors, and valves to accomplish this goal. The control system typically represents as much as 10% of the full cost of a new process plant.

With such a big investment, you might like to know if the control system is doing its job. It would seem a simple thing to monitor variability and answer that question. One could simply calculate the standard deviation of each variable every hour and determine if the control system is doing its job. But it turns out that such a simple solution is of limited value.

This paper presents seven methods to analyze controller variability. Each method provides insights into one or more aspects of control performance. The seven methods presented are:

- Basic statistics
- Scaled statistics
- Normalized measures
- Instrument performance
- Controller performance metrics
- Steady-State Analytics
- Signal Pattern Analysis

Attendees will learn how each different method works, how to apply these methods to controllers in their plants, and how to interpret the results from each method. Practical examples from industry are used to demonstrate the application of these methods using sampled date from process plant operations.

620660 The Digital Transformation Promises to Change Innovation in the Materials Industry. Will It?

Abstract:

The industry is challenged to innovate faster than ever while also delivering sustainability improvements that require profound change through the value chain. Simultaneously with that pressure, digitalization and AI are advertised as revolutions that has caused many to fear becoming obsolete. This dichotomy of extreme pressure and emerging, revolutionary technologies is causing fast, significant change in the industry and the innovation process. In this talk, I would like to share my view on what this impact will be, what risks exist, and what, in my view, will drive long-term success.

620661 Digital Twins & Smart Manufacturing –*Where Data Analytics Meet Modeling & Optimization*

Abstract:

Modeling and Optimization constitute core competencies of process systems and chemical engineering fields, and play pivotal role in the analysis, simulation, design and operational activities that we perform daily in order to improve the performance and sustainability of energy and manufacturing processes and supply chains. The re-emergence of AI and Big Data revolution that we are all witnessing at the moment, also provide an opportunity and a platform toward real-time decision-making. While often perceived as different fields, in the context of Smart Manufacturing, Data Analytics and Modeling/Optimization can in fact be viewed as 'two sides of the same coin'.

In this lecture, we will attempt to elucidate this argument through the lenses of ***hybrid modeling and digital twins***, which bring together the advantages and power of high-fidelity modeling & optimization and AI/machine learning technologies and tools, and their synergy. A central theme will be that the evolution/revolution in data-driven methods in the context of process modeling and automation is becoming a critical component in advancing the energy & manufacturing transition agenda toward clean energy, increased process/energy efficiency, sustainability and resilience, and ultimately economic and societal prosperity.

AUTHOR INDEX

AIChE
120 Wall Street, FL 23
New York, NY 10005-4020

ISBN 978-1-7138-5342-8

9 781713 853428

CURRAN ASSOCIATES INC.
proceedings
.com